GENETIC REFLECTIONS

A COLORING BOOK

by

Dr. Ahna Skop

Elif Kurt

Caitlin Marks

UW-Madison

About

Genetic Reflections: A Coloring Book aims to inspire young students and the public to explore the beauty of science and genetics. The organisms in this book are considered 'model' organisms, as they are widely studied in laboratories with hopes to understand human biology, disease pathologies, and ways to improve agricultural crops. Despite the great differences in shape and size, on the genetic level there are lots of similarities.

In every species, DNA sequences consist of the same four building blocks (G, C, A, & T). However, slight changes in their use, even in the same gene, can occur in each species. The way our bodies and cells work are well conserved throughout evolution, even in species that may look very different from us. The beauty of our world, even on the cellular level, is apparent.

Genetic Reflections: A Coloring Book is a collaboration between Ahna Skop, Elif Kurt and Caitlin Marks; two UW-Madison undergraduate Skop Lab members. This coloring book is the outcome of a year-long independent study in Life Sciences Communication with goals to broadly disseminate the *Genetic Reflections* scientific glass art installation created by Angela Johnson and Ahna Skop.

Visit

The public is welcome to visit and engage with the *Genetic Reflections* scientific art piece inside the UW-Madison Biotech Center at 425 Henry Mall, Madison, WI 53706. For large groups, field trips can be scheduled via BioTrek (biotech.wisc.edu/biotrek).

Acknowledgements

We would like to express our appreciation to Dr. Diana Chu, Dr. Tom Zinnen, Liz Jesse, and Nathaniel Sharp for valuable and constructive suggestions during the course of creating this coloring book. We would like to thank the Life Sciences Communication department for supporting this independent student project. Lastly, the *Genetic Reflections* scientific art piece and this associated coloring book would have not occurred without the generous support from the National Science Foundation (NSF,MCB-1716298), UW-Madison Biotech Center, the Wisconsin Materials Research Science and Engineering Center (MRSEC) (NSF, DMR-1720415), and the UW-Madison Laboratory of Genetics. Part of the proceeds of this book will be donated to charities and programs that support STEAM (Science, Technology, Engineering, Arts, and Mathematics) educational innovations or public outreach events.

Dr. Ahna Skop is an internationally known geneticist, artist, science communicator, and champion for the underrepresented in science. She studies how cells divide and communicate, knowing that failures in this process can lead to cancer, microcephaly and neurodegenerative disease. Cell division is highly dependent on visual data, which dovetails perfectly with one of her other passions, art.

Ahna has curated and contributed to several scientific art installations on the UW-Madison campus and internationally, including *Genetic Reflections* for which this coloring book is associated. Science and art are necessary for innovation in science, and the melding of these two fields have proven to be excellent ways to encourage young students to pursue careers in science, for which Ahna is equally passionate about.

Ahna is genetically an artist. Her father, Michael Skop, was a classically trained fine artist who studied with Mestrovic (a pupil of Rodin), and also taught anatomy to medical students. Her father operated an art school at their home studio for over 30 years, which attracted artists, musicians, and philosophers from all over the world. Her mother was a high school art educator, ceramicist, and has dabbled in fiber art, sculpture and painting. Her two sisters (Tarsia and Zesha) and brother (Damien) are outstanding graphic and industrial designers.

Ahna is currently a Professor in the Laboratory of Genetics, and also an affiliate faculty member in Life Sciences Communication and the Division of the Arts. She majored in biology and minored in ceramics at Syracuse University (1990-1994). She received her Ph.D. in Cell and Molecular Biology at the University of Wisconsin-Madison (1994-2000), and conducted her post-doctoral work at UC-Berkeley (2000-2003). She was named as a AAAS Remarkable Woman in Science in 2008, a Kavli Fellow by the National Academy of Sciences in 2015, and received the prestigious Presidential Early Career Award for Scientists and Engineers (PECASE) in 2006. More recently, she was named an IF/THEN Ambassadors for Women in STEM by the AAAS in 2019, which aims to serve as high-profile role models for middle school girls. One of her great hobbies is traveling, where she partakes in local cooking classes, and she enjoys cooking and baking, including scientific cakes, all while she manages a foodblog, foodskop.com in her free time.

Email: skop@wisc.edu
Twitter: @foodskop
Instagram: @foodskop
Facebook: https://www.facebook.com/ahnaskop/
LinkedIn: https://www.linkedin.com/in/ahnaskop/

Elif Kurt is a recent UW-Madison graduate who majored in Genetics and Genomics. She worked on Genetic Reflections: A Coloring Book as part of her Independent Study work in Life Sciences Communication (LSC 299) during her senior year. Elif plans on going to medical school and currently volunteers at the UW-Madison hospital. Elif also tutors middle and high school students using art to teach scientific topics that normally appear daunting. Elif hopes to inspire others to get involved with science and art in their own schools and communities. Elif enjoys spending time with her cat, practicing mindful thinking and meditation, and illustrating in her free time.

Email: elifmkurt@gmail.com
Twitter: @digital_elif
Instagram: @digital.elif
LinkedIn: https://www.linkedin.com/in/elifmkurt/

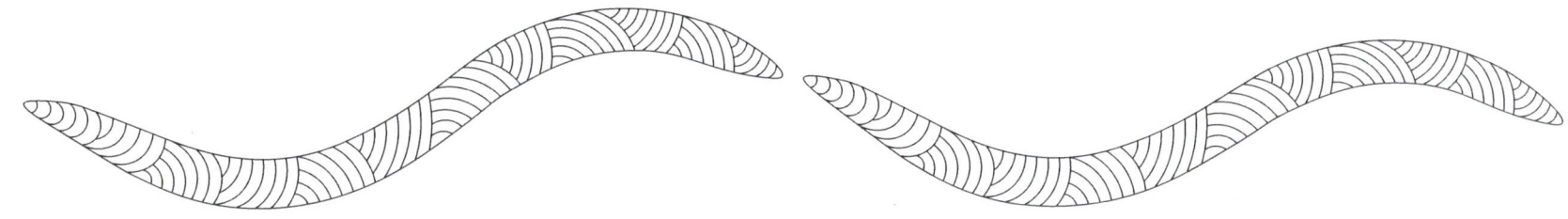

Caitlin Marks is a recent UW-Madison graduate who majored in Genetics and Genomics and a certificate in Studio Arts. She worked on Genetic Reflections: A Coloring Book as part of her Independent Study work in Life Sciences Communication (LSC 299) during her senior year. Caitlin has plans to get her Ph.D. by focusing on infectious diseases, in hopes to someday work at the Center for Disease Control (CDC) or the World Health Organization (WHO). Caitlin served as the Vice-President of the Microbiology Club at UW-Madison, where she has encouraged other students to participate in science outreach. Caitlin likes to spend her free time hiking and backpacking in Wisconsin, and also national parks, and is an avid trivia player. Caitlin also designs T-shirts and signs for weddings and events.

Email: marks7@wisc.edu
Instagram: @helianthus_designs
LinkedIn: https://www.linkedin.com/in/caitlinmarks/

Text © 2020 by Dr. Ahna Skop

Illustrations © 2020 by Elif Kurt and Caitlin Marks

All rights reserved. No part of this book may be used or reproduced
in any manner whatsover without written permission,
except in the case of brief quotations in a book review.
For more information, address: skop@wisc.edu

Published 2020

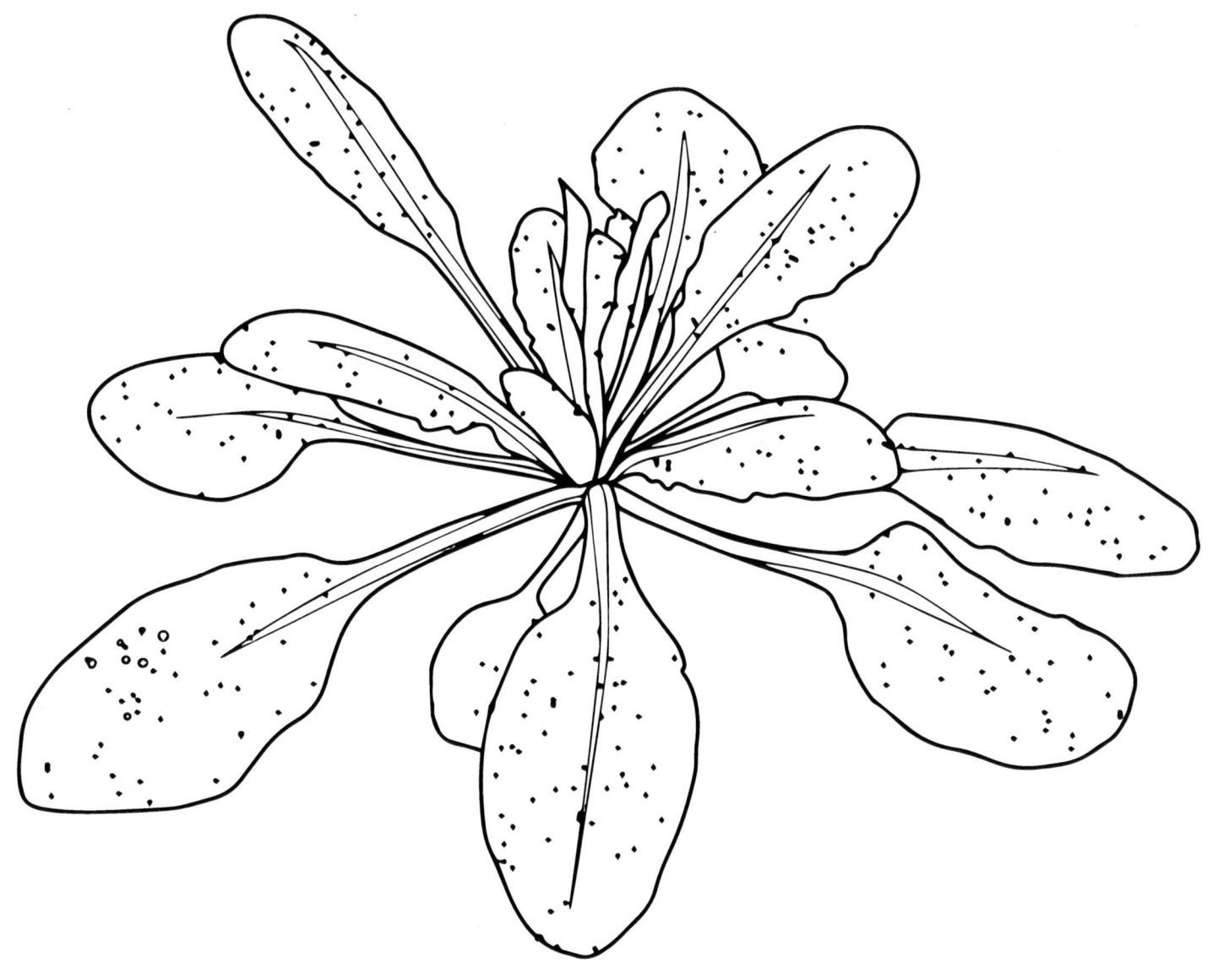

A is for *Arabidopsis thaliana*

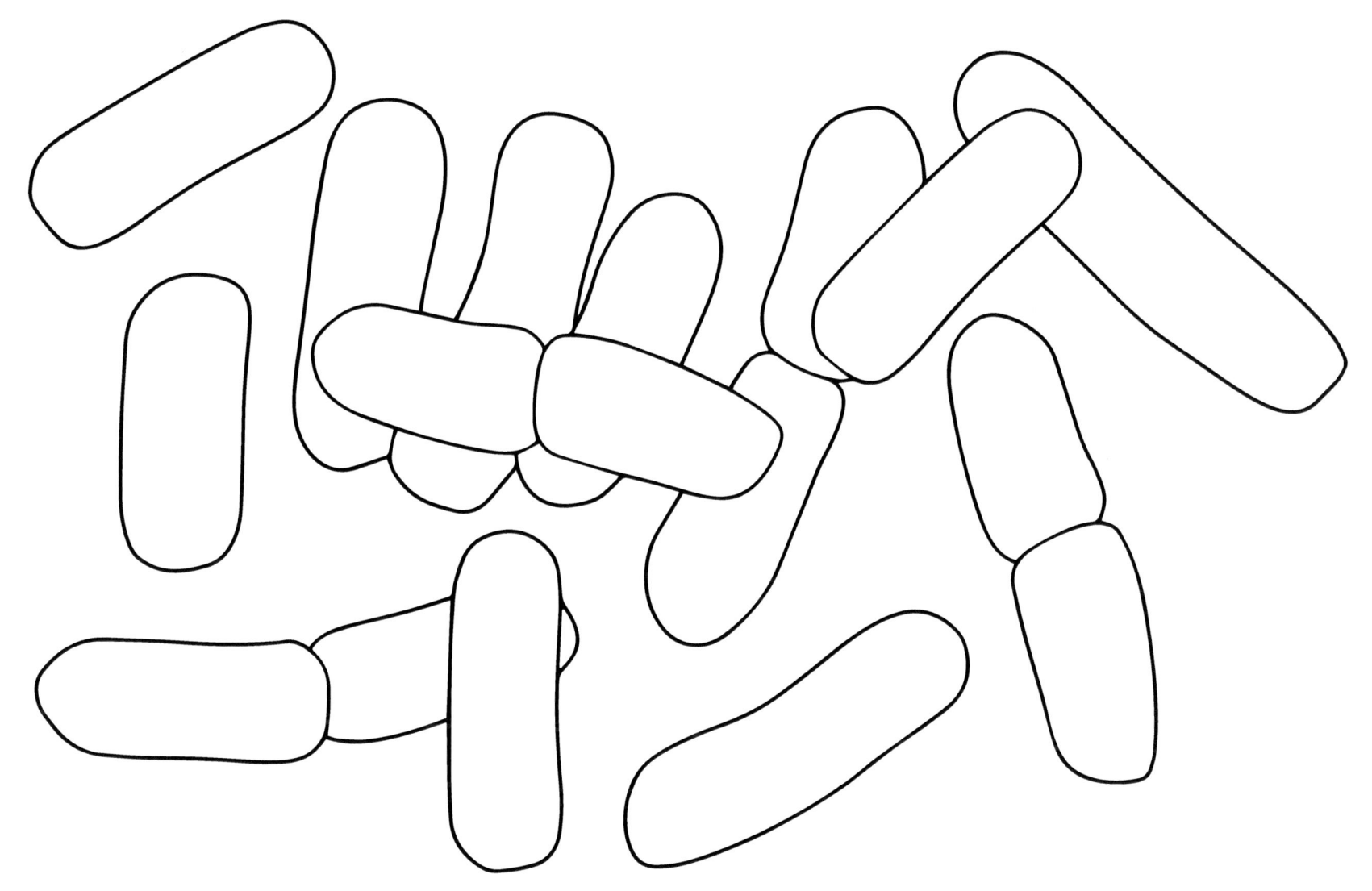

B is for Bacteria

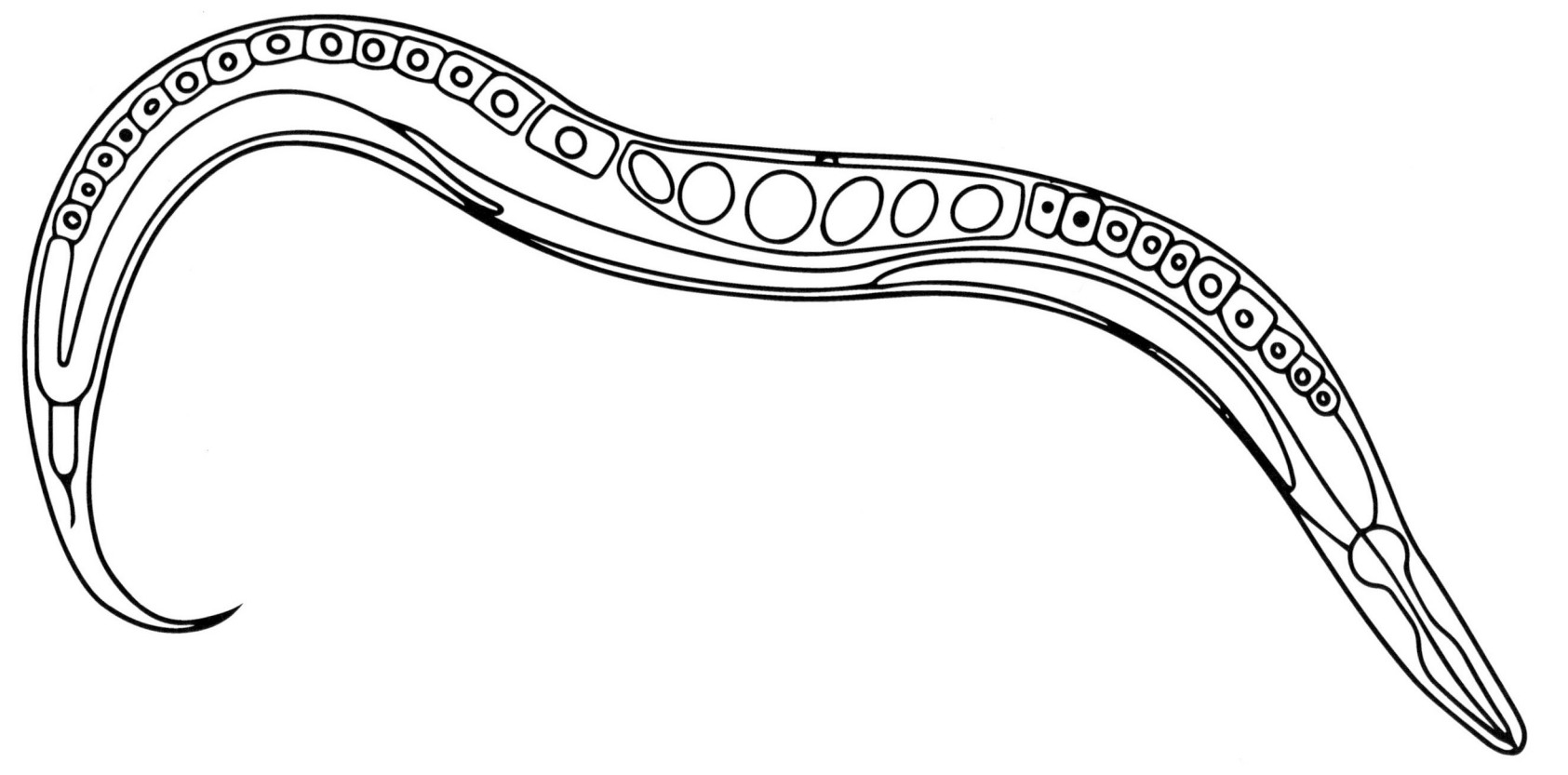

C is for *Caenorhabditis elegans*

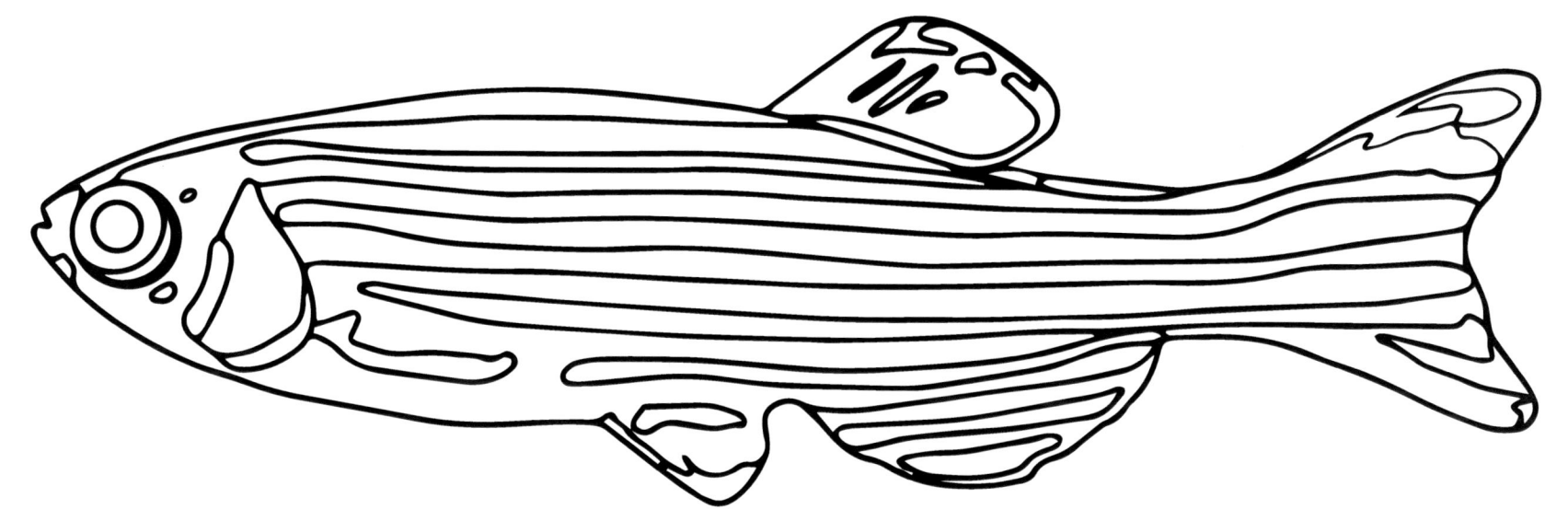

D is for *Danio rerio*

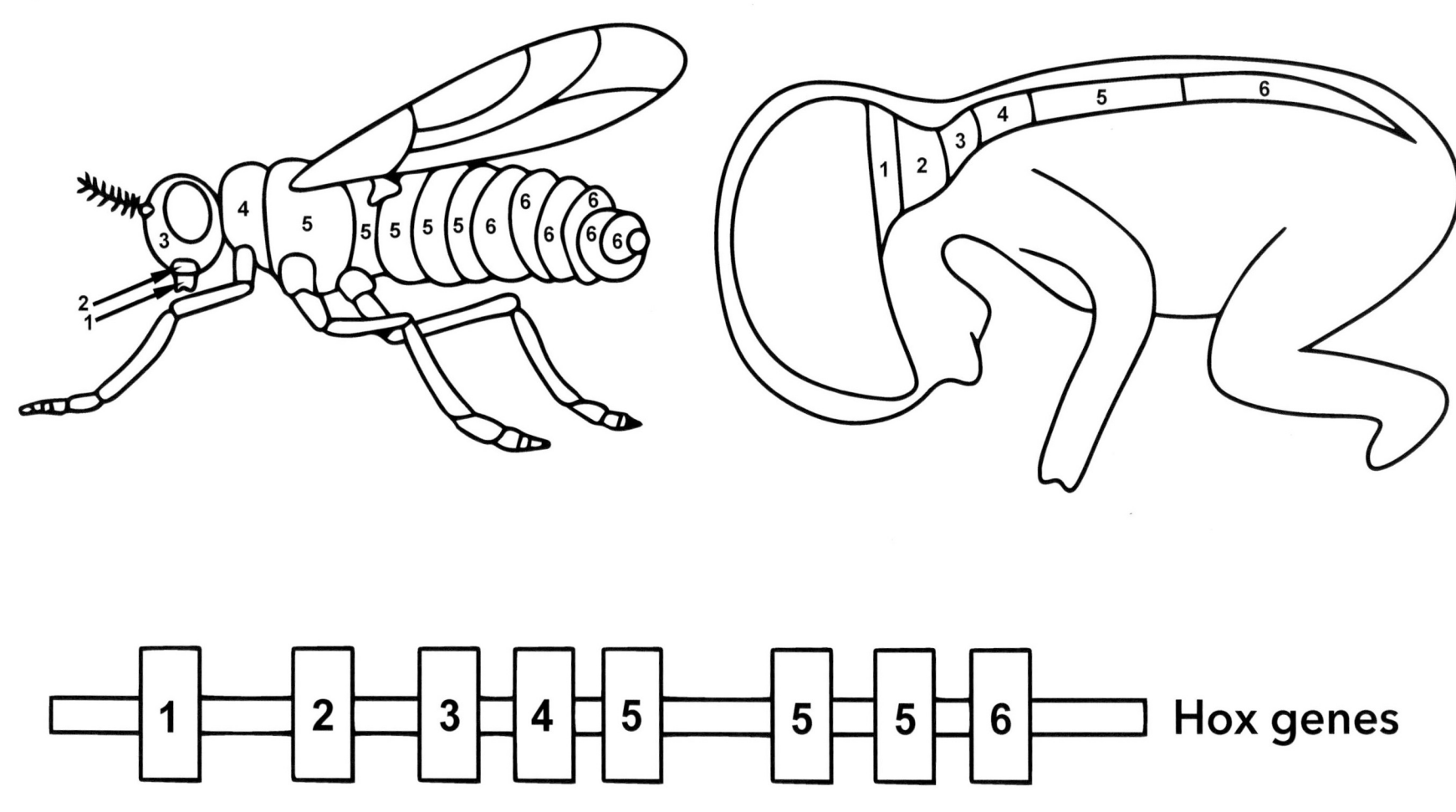

E is for Evolution

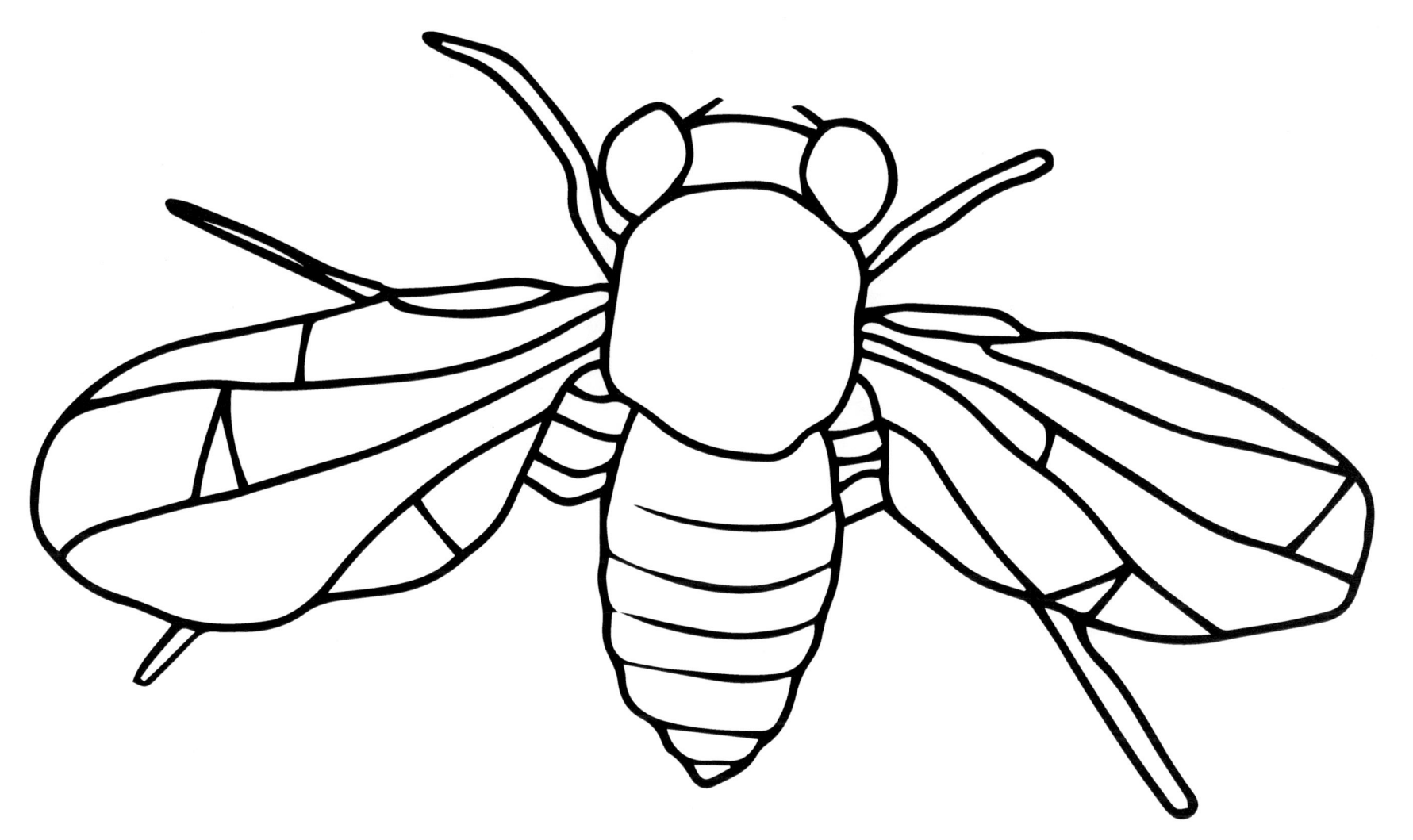

F is for Fruit fly

G is for Genetic sequence

H is for Human

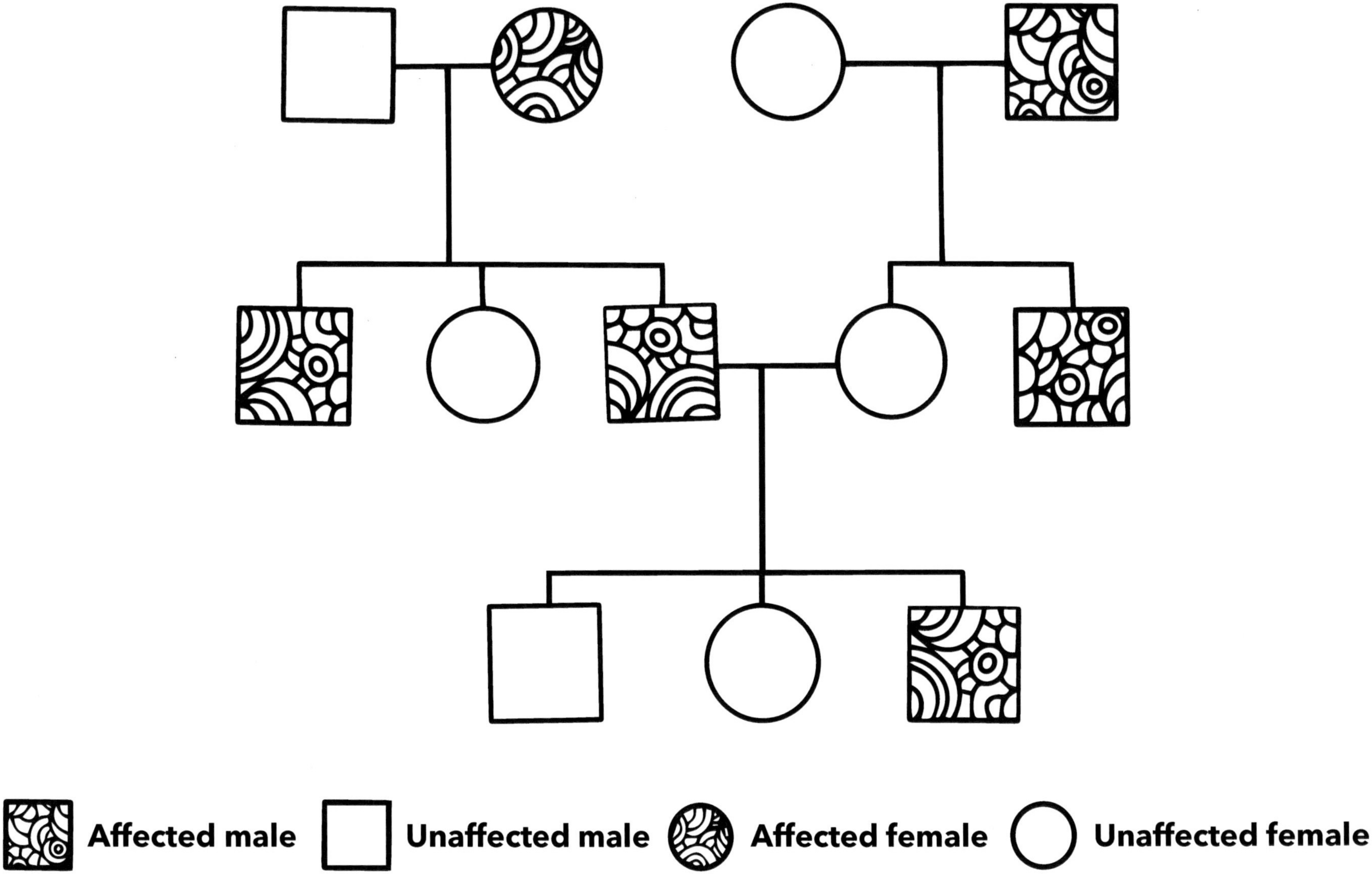

I is for Inheritance

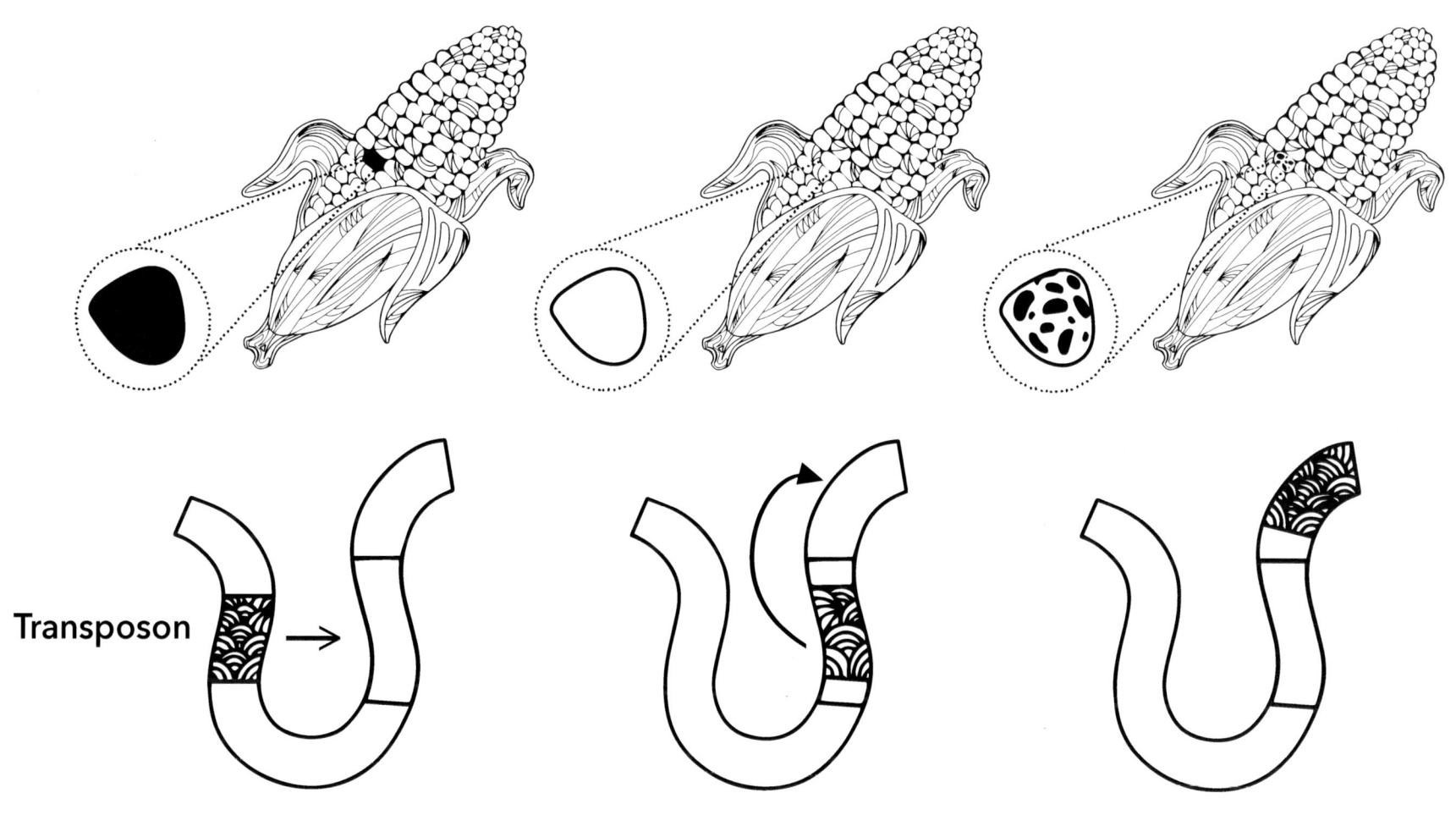

J is for Jumping gene

Wild-type

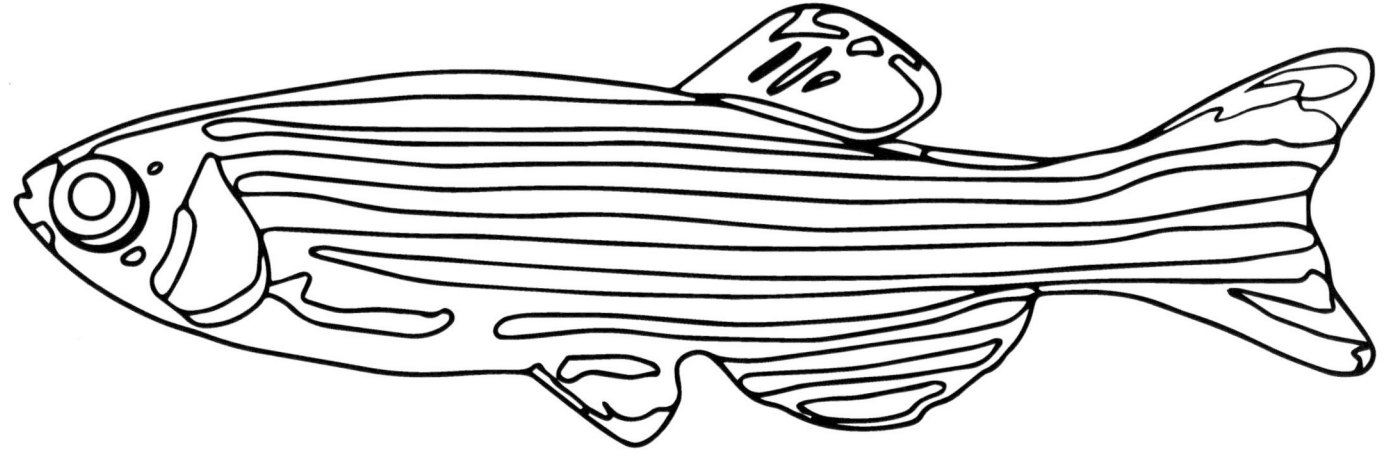

nacre mutant

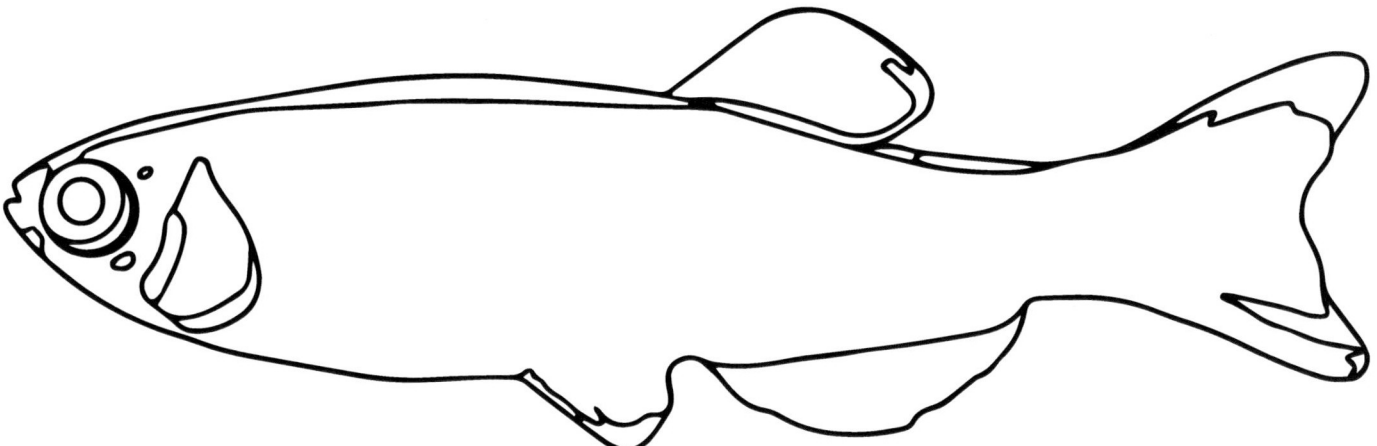

K is for Knockout mutant

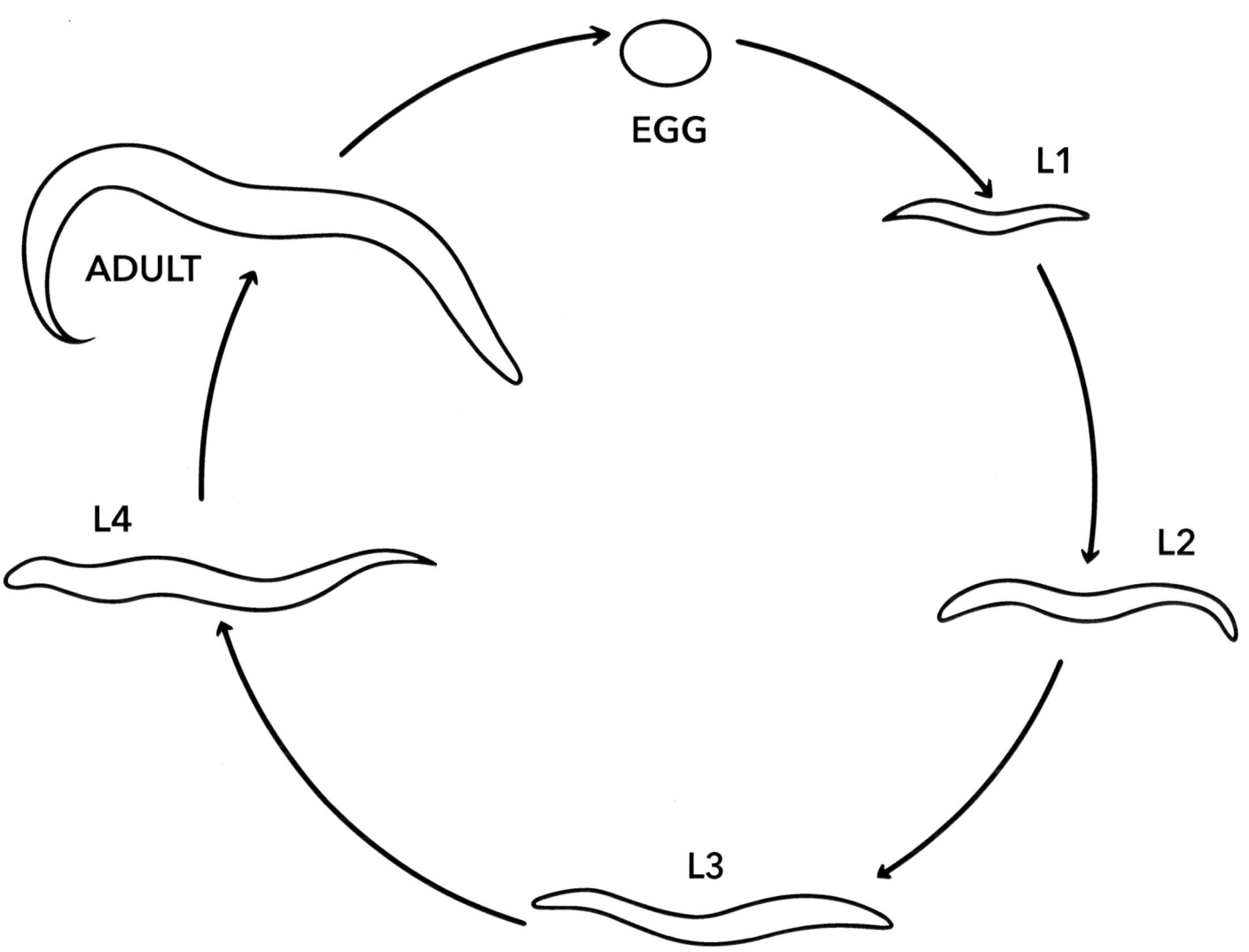

M is for Mus musculus

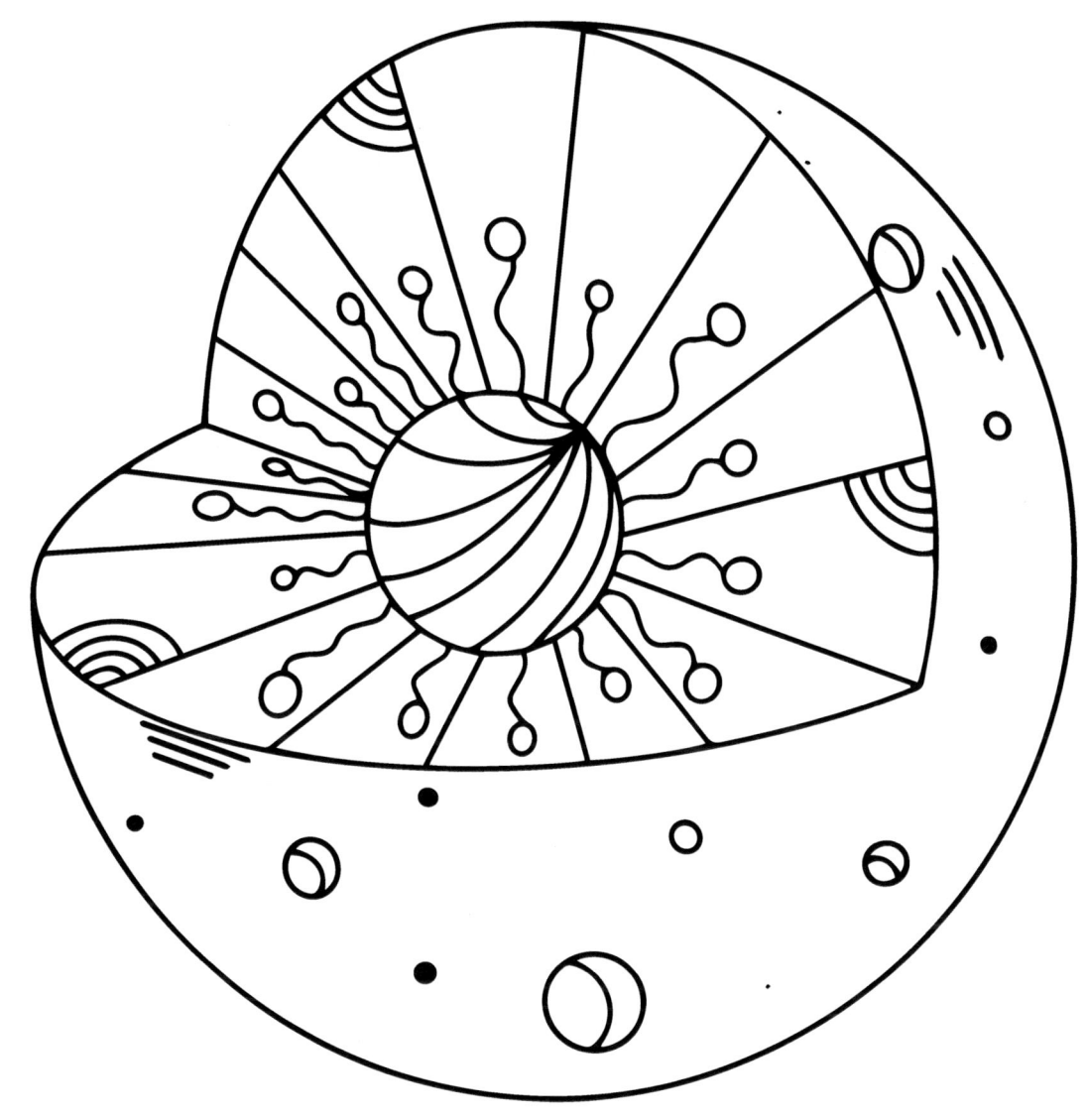

N is for Nucleus

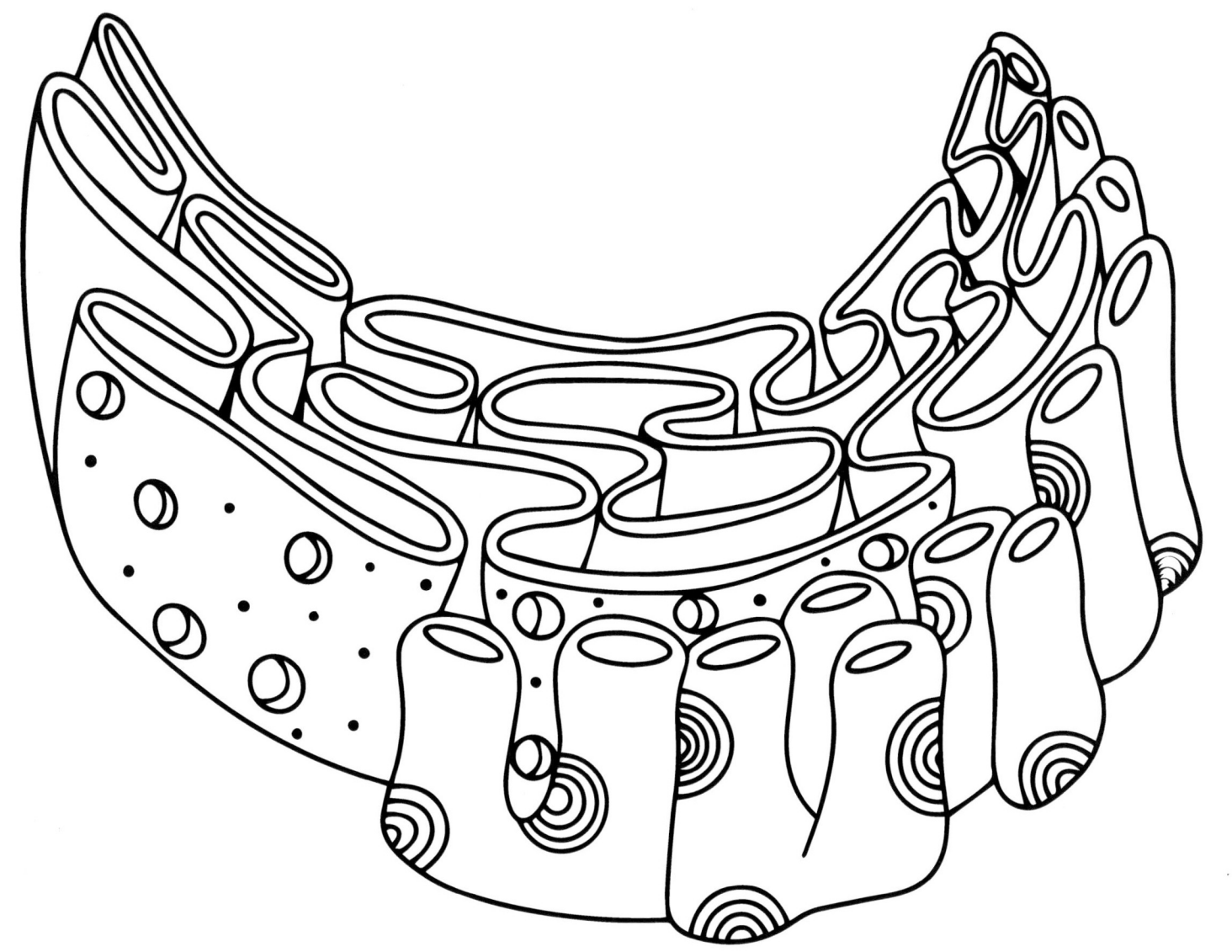

O is for Organelle

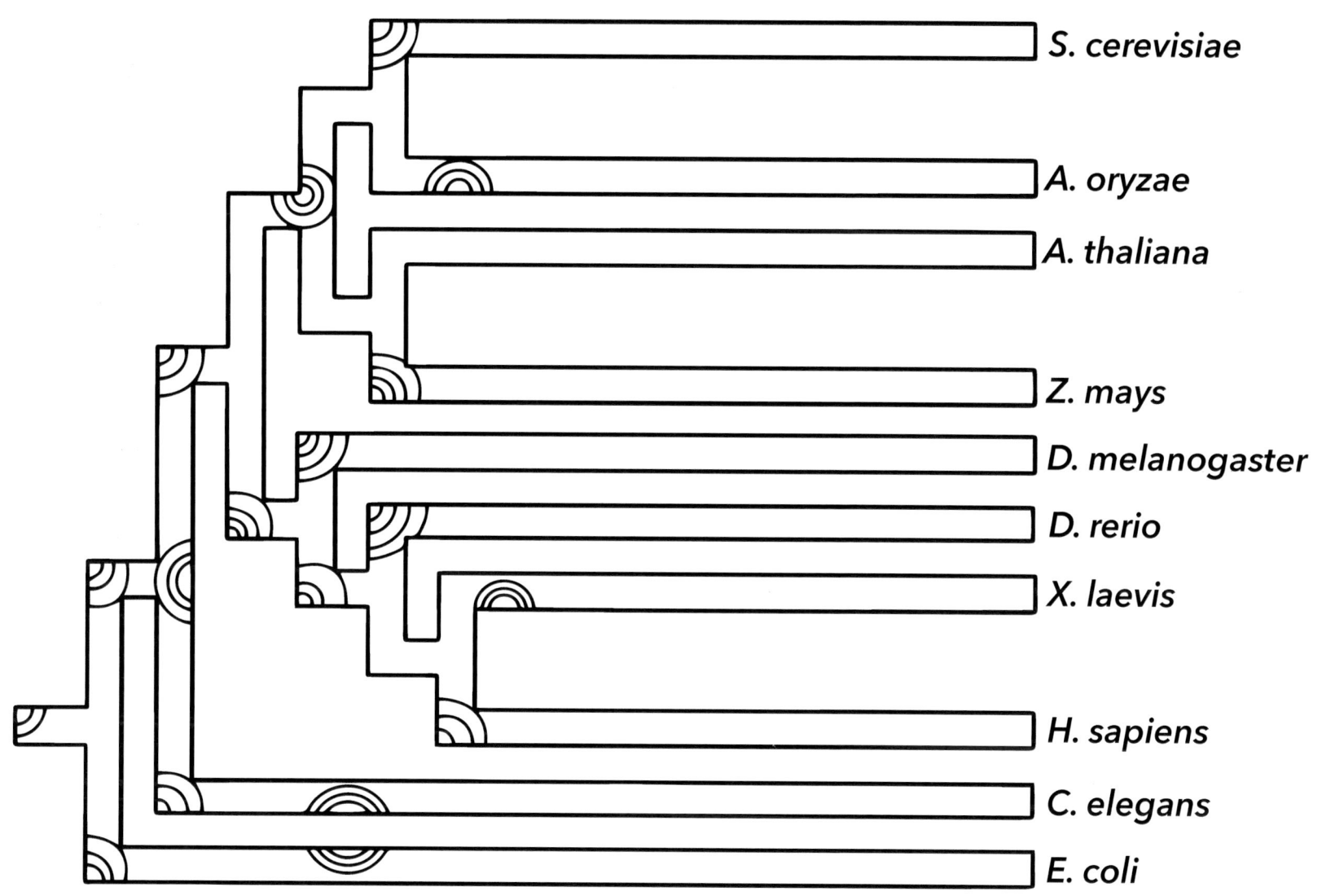

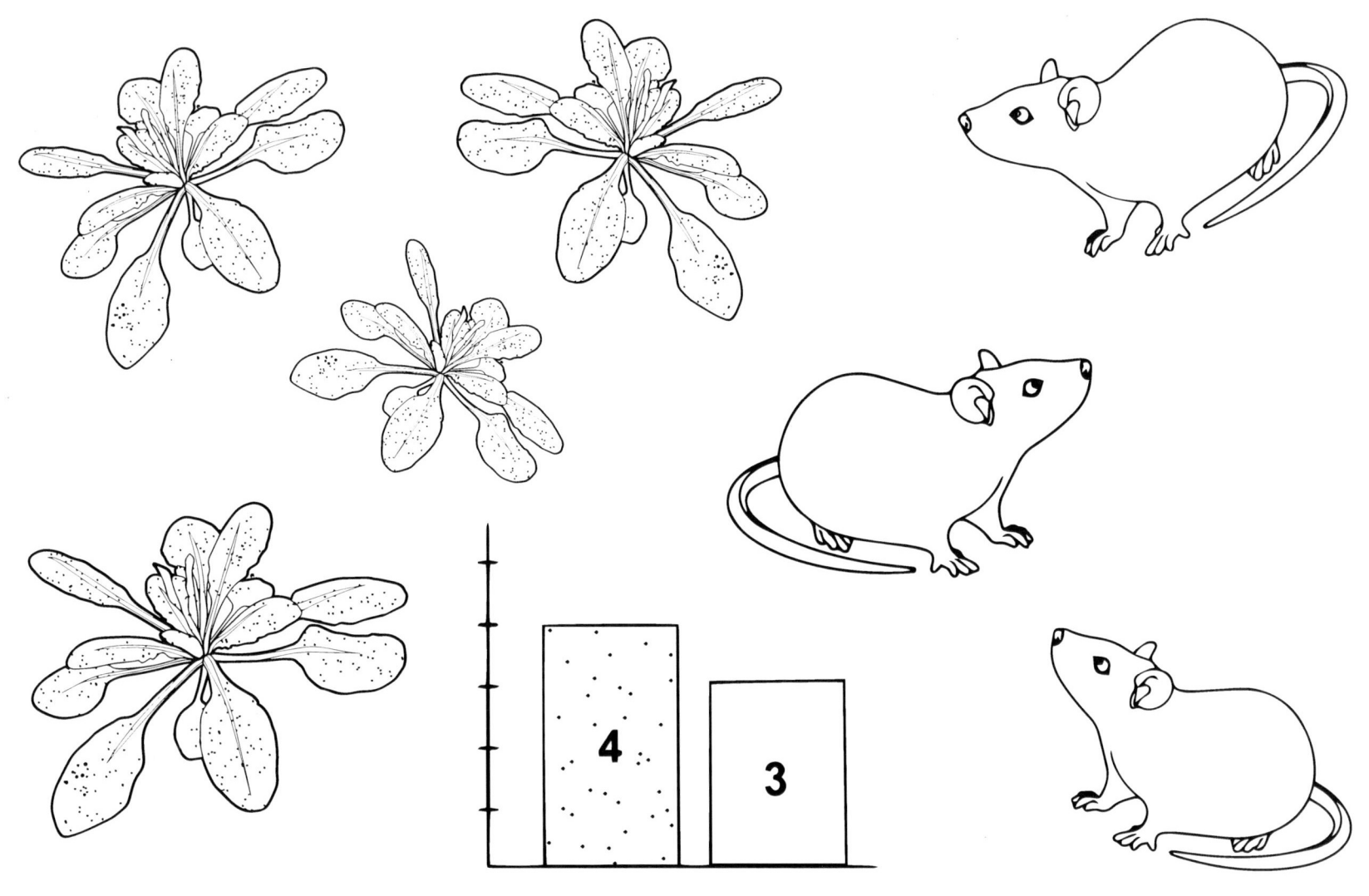

Q is for Quantify

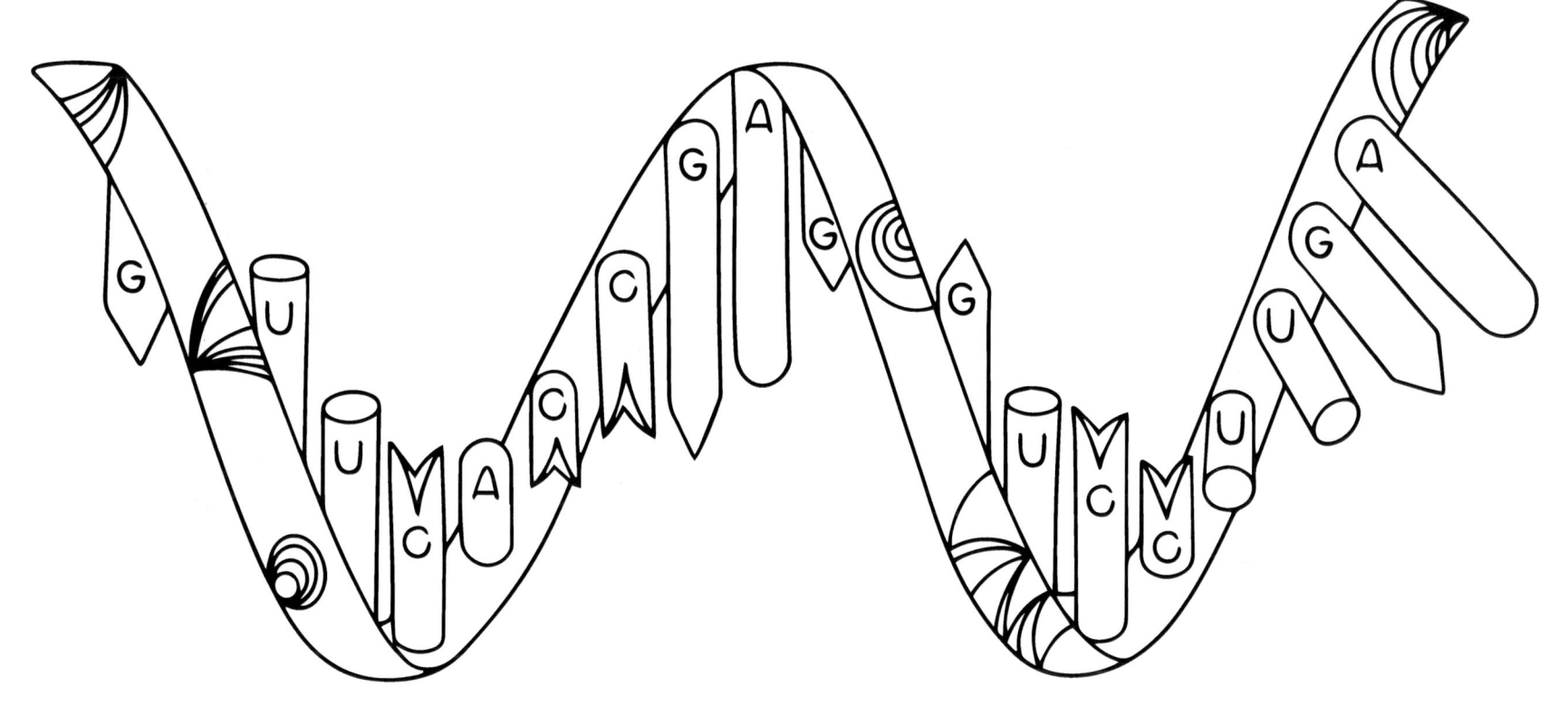

R is for RNA

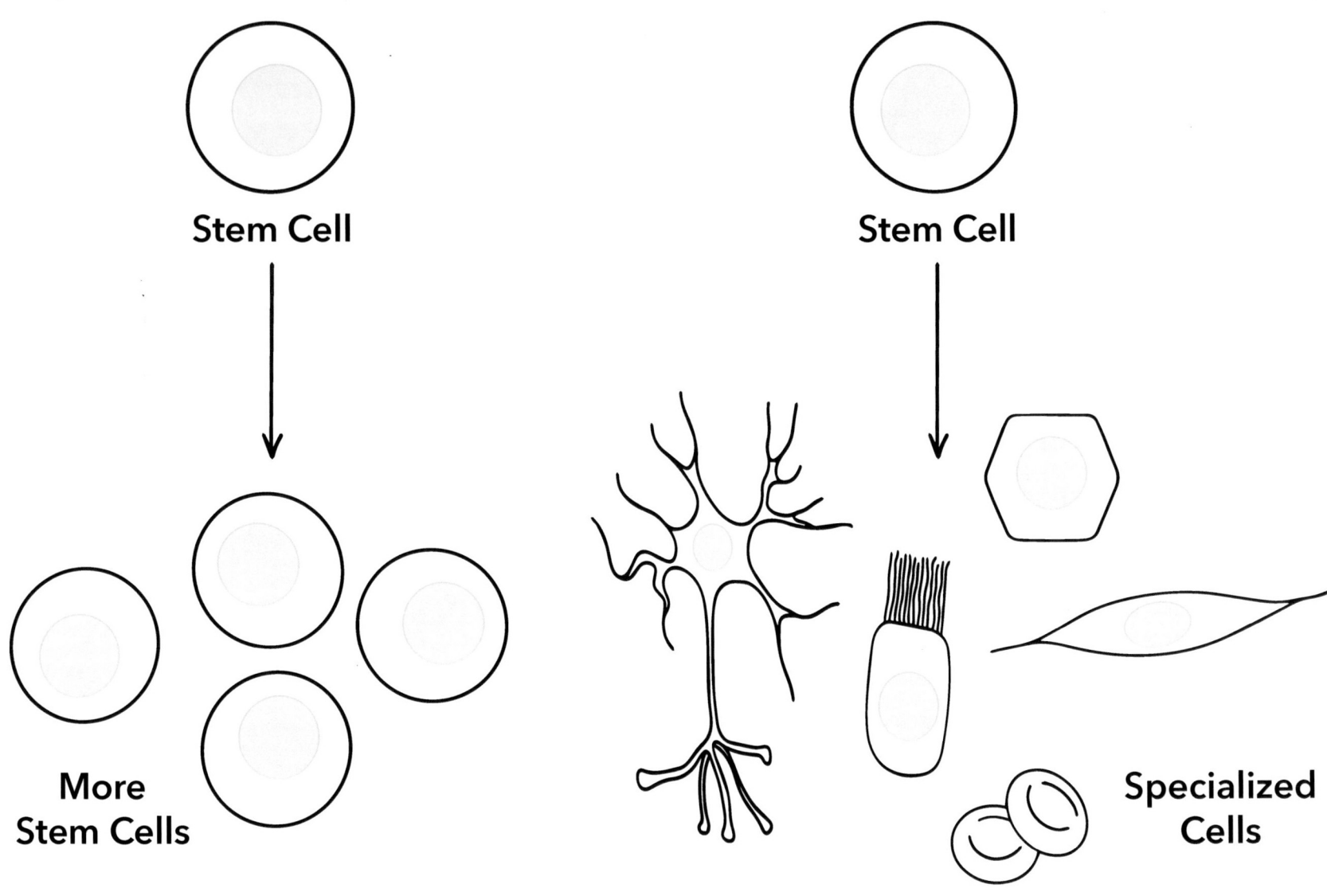

S is for Stem cells

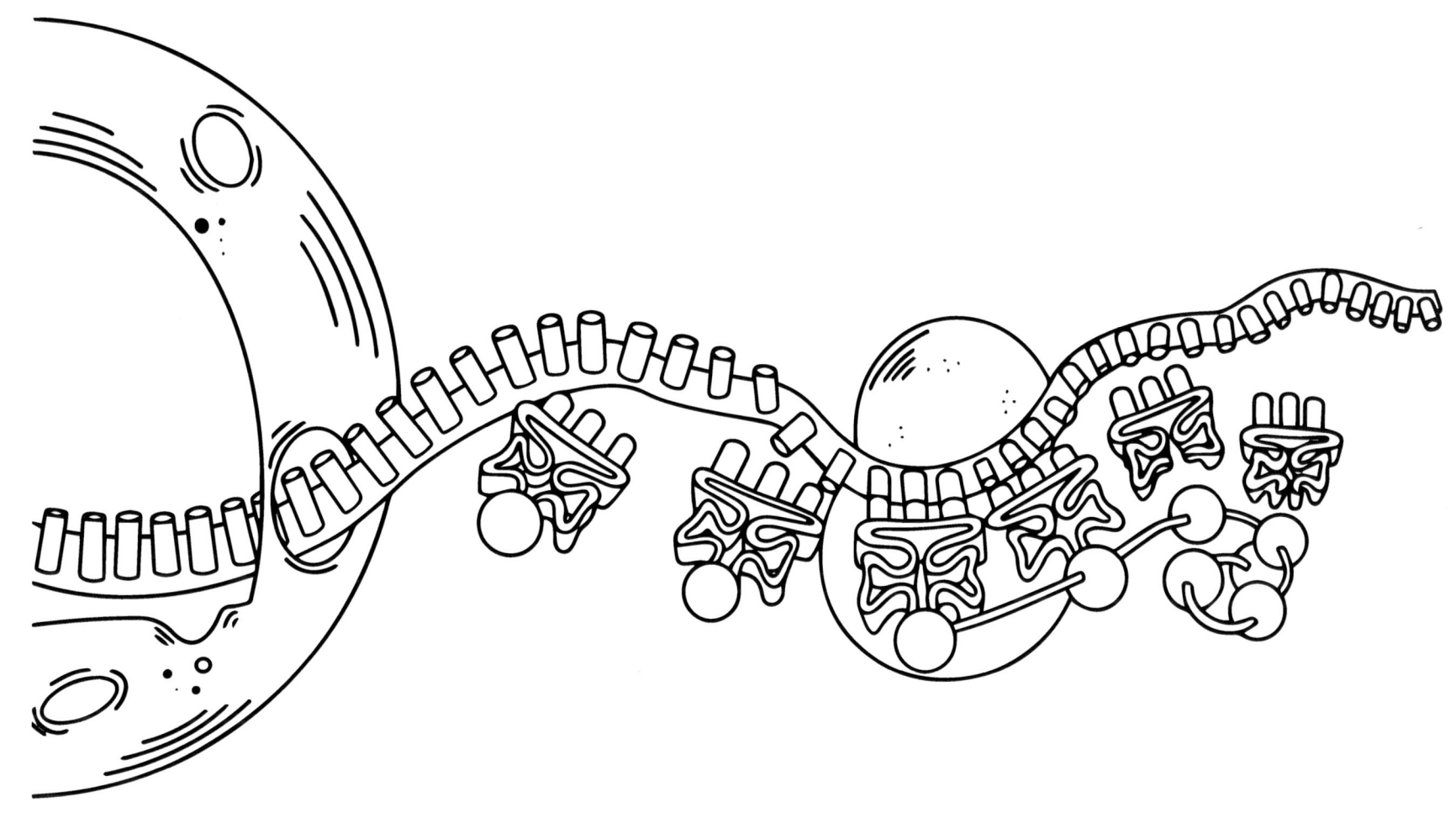

T is for Translation

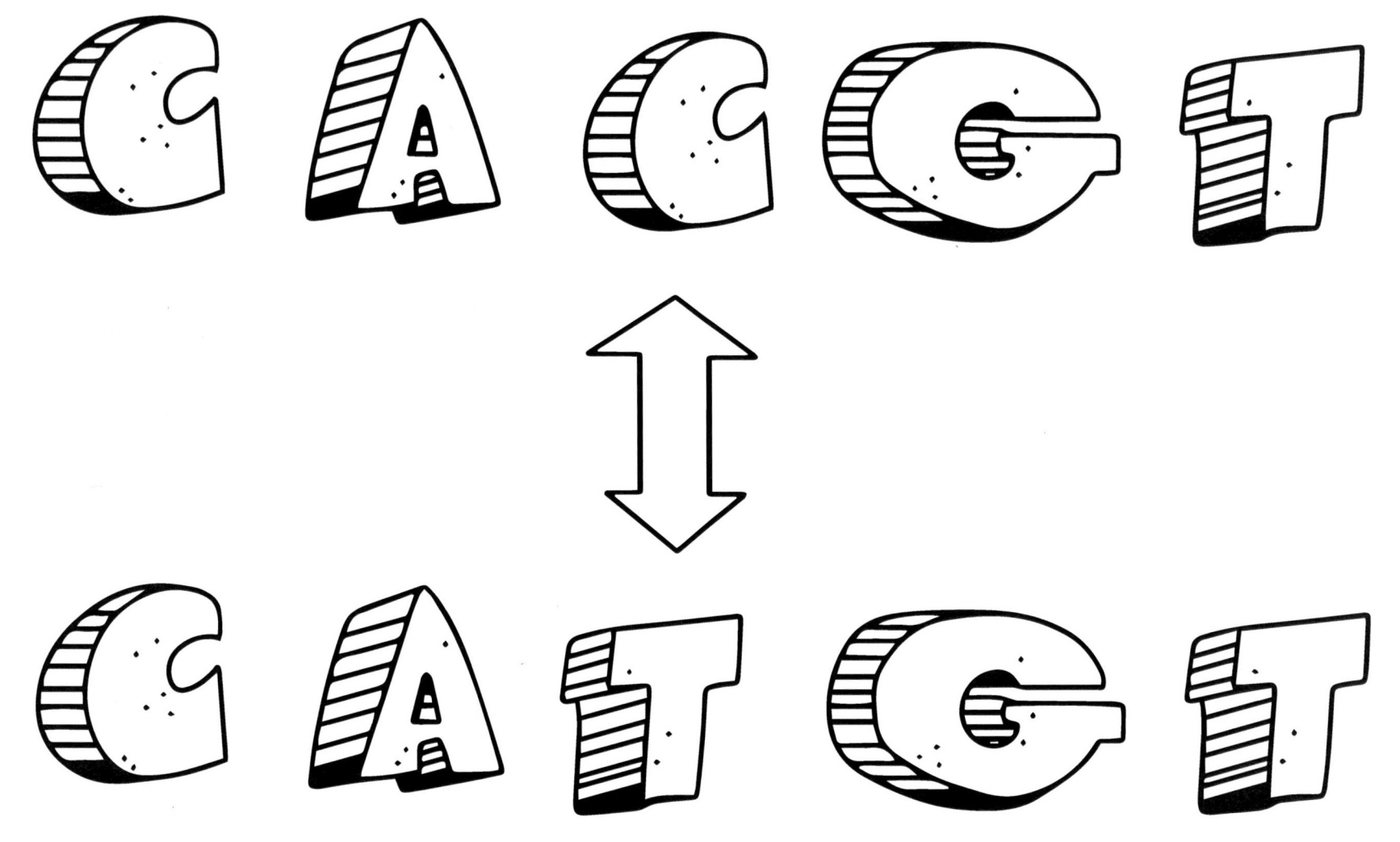

U is for Unique DNA sequence

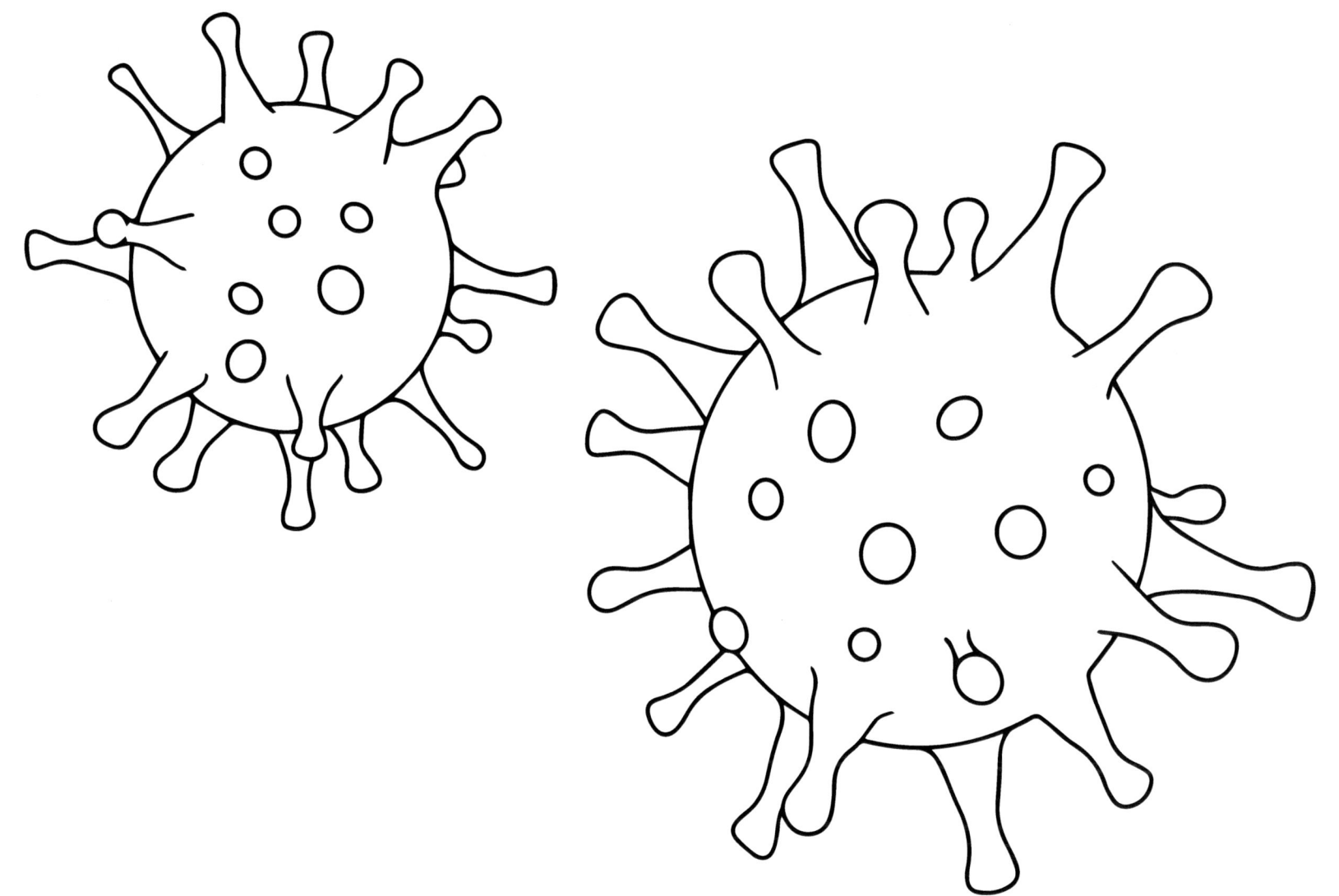

V is for Virus

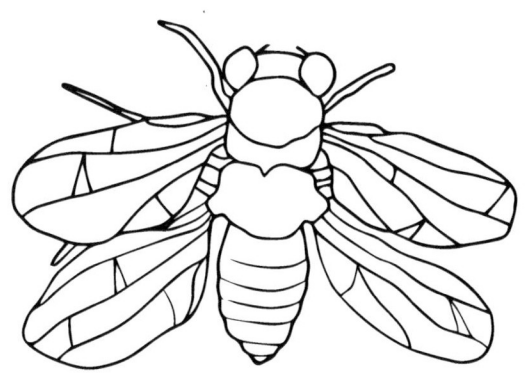
bithorax mutant

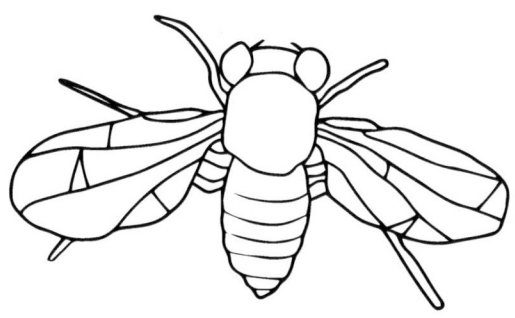
Wild-type

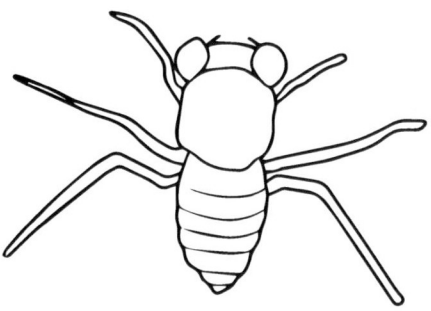
wingless mutant

W is for Wild-type

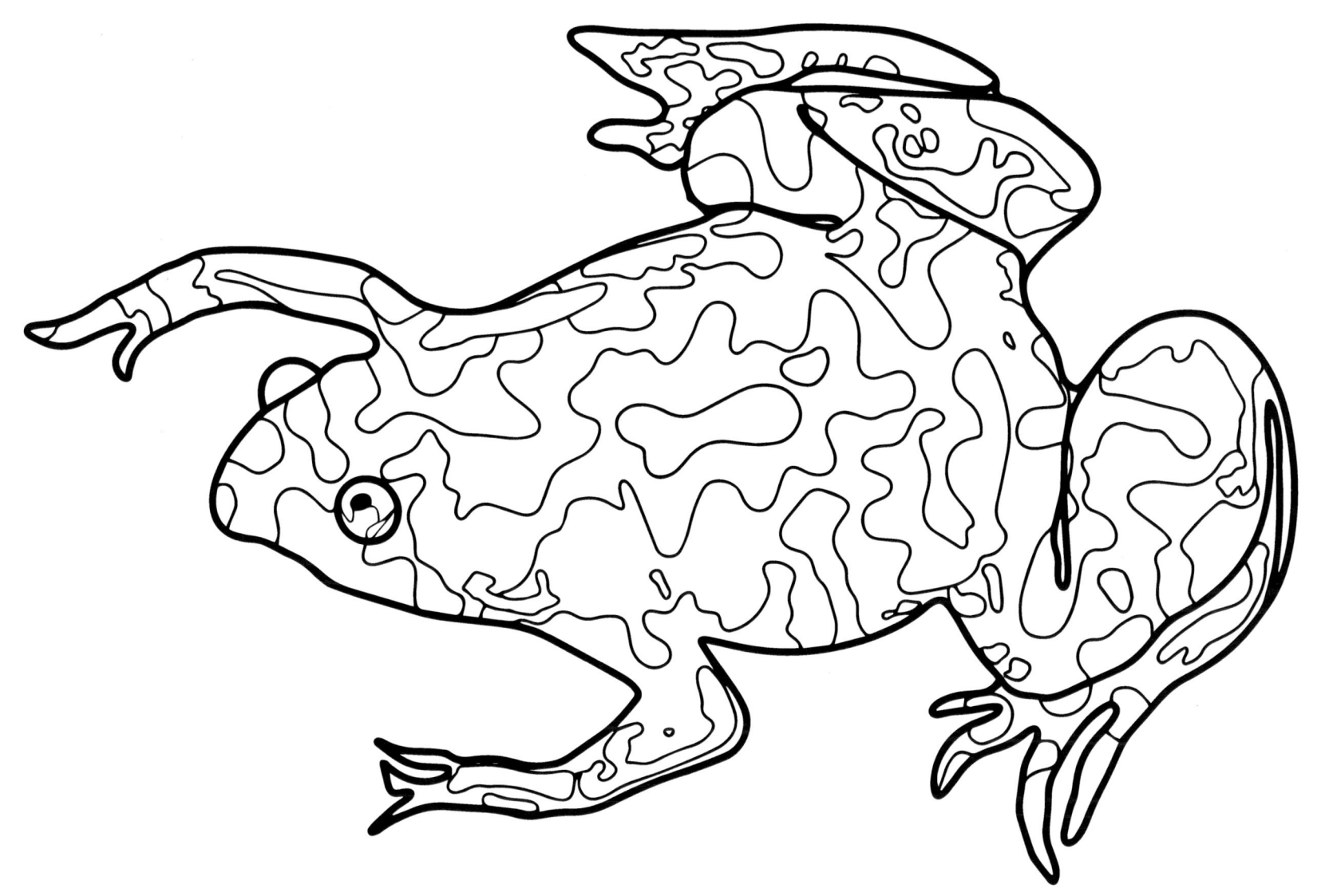

X is for Xenopus tropicalis

Y is for Yeast

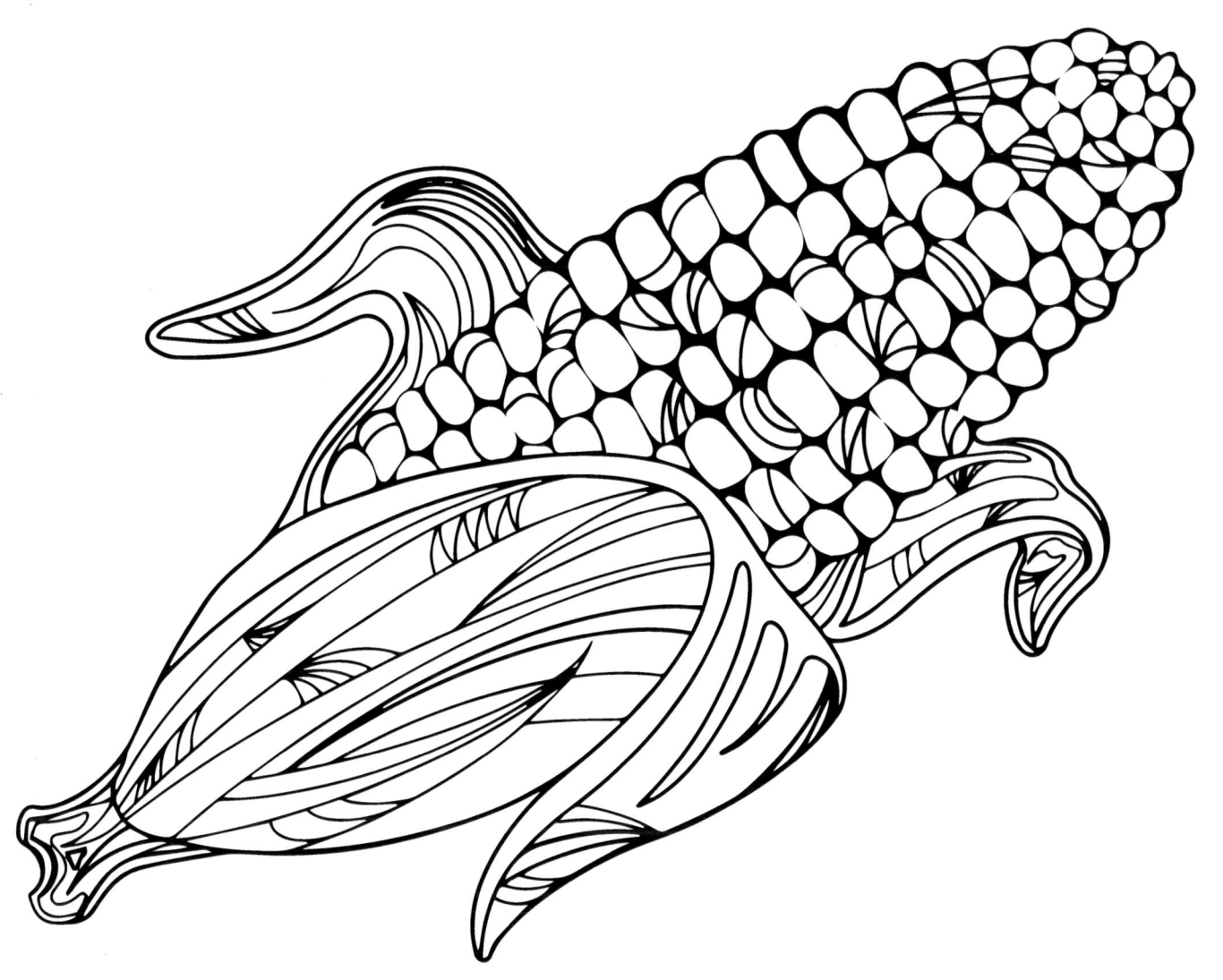

Z is for *Zea mays*

Glossary

Arabidopsis thaliana: is a type of mustard plant native to Europe, Asia and Africa and a popular model organism in genetics. It was the first plant to have its genome sequenced and has helped us understand growth of plants and the development of many of our food crops.

Bacteria: are microscopic, single-celled organisms that thrive in diverse environments such as soil, the ocean, and inside the human gut. Sometimes bacteria lend us a helping hand, such as by curdling milk into yogurt or helping us with our digestion. Bacteria can also be devastating to human health, causing diseases like pneumonia.

Bithorax mutant: is a fruit fly (Drosophila melanogaster) mutant where the entire section of the body (thorax) is duplicated. This leads to a fly with four wings instead of two. First isolated in 1915, by Calvin Bridges, it was later studied by Edward B. Lewis, who found out these genes code for proteins that control the development of the thorax and abdomen or gut of the animal, for which Lewis won the Nobel Prize in 1995.

Caenorhabditis elegans: is a transparent species of roundworm or nematode that feeds on bacteria living on rotting fruit. *C. elegans* is a powerful model organism in genetics, particularly for its use to understand neurobiology, aging and cell biology. It was the first animal to have its genome sequenced. Its entire nervous system was mapped in exquisite detail by John G. White (Dr. Ahna Skop's Ph.D. mentor). In 2002, Sydney Brenner, Robert Horvitz, and John Sulston won the Nobel Prize in Medicine for their work on cell division and cell death using *C. elegans*.

Danio rerio: or zebrafish, is a freshwater fish native to the Himalayan region and belonging to the minnow family. The zebrafish has a backbone and is widely used as a vertebrate model organism in genetics. It is particularly prized for its transparency, embryonic development, regenerative abilities of the heart, and a simple cardiovascular system. George Streisinger, a tropic fish enthusiast at the University of Oregon, was looking for a vertebrate model that was less complex and cheaper to rear than the mouse (*Mus musculus*). The zebrafish model system combines many of the best features of all of the model systems, like transparency of the embryos, external development, and short generation time.

Evolution: is a scientific theory that explains how living things change over time, and how thing come to be the way they are. Genetic researchers often compare DNA sequences between organisms or to determine how similar they are to each other. Charles Darwin in his book, "On the Origin of Species" in 1859 put forward much evidence that evolution had occurred. He proposed that natural selection was the way evolution had taken place.

Hox Genes: are a set of genes whose proteins direct the positional cues for the development of an organism. For example, one gene can encode where your arms will be placed along your body axis. Mutations in these genes often lead to extra limbs. The Hox genes, Bithorax, were first discovered by Calvin Bridges in Thomas Hunt Morgan's laboratory in 1915.

Fruit fly: is a small fly that feeds on rotting fruits and vegetables. This fruit fly, or *Drosophila melanogaster*, is a popular genetic model organism widely used to study behavior, brain function, molecular biology, evolutionary and population biology, and embryonic development. In 1995, Christiane Nüsslein-Volhard, Eric Weichaus and Edward Lewis shared the 1995 Nobel Prize for their pioneering genetic screens that revealed the genetic control of early embryonic development in *Drosophila*. Nüsslein-Volhard also went on to help establish zebrafish as a model organism to study vertebrate development.

Genetic sequence: is the order of the nucleotides (G, C, A or T) in DNA.

Human being: are primate mammals, especially *Homo sapiens*, who are distinguished by a more highly developed brain and capacity for speech and abstract reasoning.

Inheritance: is when a child receives traits or behaviors, or characteristics from their parents. Your parents genes largely determine what you will be like. For example, eye color or a genetic disease is influenced by your parents DNA.

Jumping genes: is a sequence of DNA that can move to a new position in the genome of the same cell. They are more often called transposons and were first discovered by Barbara McClintock while working on maize or corn. She received a Nobel Prize for her work in 1983.

Knockout mutant: is where scientists have turned off or "knocked out" a gene in an organism. One technique used to knock out genes is CRISPR, which is a way to edit a DNA sequence genetically. Knockout techniques are commonly used to study how a gene normally works. Knockout mutants are used with many of the model organisms in this coloring book to help us understand human disease.

Life cycle: are the stages a living organism goes through during its life. One example we show in the book is the life cycle of the nematode, *C. elegans*, from egg or embryo to the adult worm, similar to how caterpillars transform into butterflies later on in life.

Mus musculus: is a common house mouse used in genetic research to model human disease. Almost every single human gene has a counterpart in the mouse, with similar DNA and functions. Mice are useful for studying aging, neurodegenerative disease, obesity, stress, immunity, cancer, circadian rhythm and evolution, for example.

Nacre mutant: is a zebrafish (or *Danio rerio*) mutant that is commonly used in genetic research because the pigments or melanocytes in their scales are missing. Geneticists use these transparent mutants to easily spot mutated and GFP-tagged fish to understand development and the heart or brain function on live fish using high-powered microscopes.

Nucleus: is perhaps the most important large structure inside the cell. It contains nearly all of the cell's DNA. It works like the brain for the cell.

Organelle: is a part of the cell that performs a certain job, like a small machine. The illustration in this book is part of the Golgi apparatus that helps our cells process proteins, often putting sugars on them, so that they will be transported elsewhere. Think of the Golgi apparatus like the post-office of the cell.

Phylogeny: is the evolutionary history of a set of particular organisms or species. The main purpose is to determine how species (whether living or extinct) are related by ancestry or relatedness.

Quantify: is to find or calculate the amount of something. Geneticists always need to quantify things to determine how common a certain trait or mutation occurs.

RNA: or ribonucleic acid, is a cousin of DNA that encodes for proteins. RNA is often the only genetic material or nuclei acid found in some viruses, like the coronavirus or HIV. When these types of RNA viruses infect a cell, they replicate their RNA genome using host or human cell machinery. This allows the virus to then infect other cells.

Stem cells: embryonic stem cells are cells with the unique ability to develop into any cell type in the body. Stem cells provide new cells for the body as it grows and can replace cells that can become damaged or lost. They have a unique property in that they can divide over and over again to produce new stem cells or turn into specialized cells like a neuron, for example.

Translation: is the process when RNA is used as a kind of code to string together amino acids into proteins. There are about 20 different amino acids, and just as stringing together letters in different orders can give different words, stringing togother amino acids in different orders gives different proteins. Translation occurs at an organelle called the ribosome.

Unique DNA sequence: A unique DNA sequence does not appear anywhere else in the genome of an organism.

Viruses: are non-living particles that contain genetic material (DNA or RNA) that can get inside cells and make many copies of the virus. In humans viruses can often make people sick. Viruses can hide out in our cells, where medicines often cannot attack them, making them difficult to treat. Outside of our cells, most types of viruses cannot last long, but some can survive in the air, and water, or on your skin. Washing your hands with soap can kill viruses easily because soap breaks them down. Scientists have developed vaccines, which help our body build up our own antibodies which act as defenses against the virus. Vaccines often prevent people from being attacked by the same strain of virus again.

Wild-Type: is the most common, or laboratory form of an organism used in genetic research. It refers to the typical form of a species as it occurs in nature.

Wingless mutant: is a fruit fly (Drosophila melanogaster) mutant missing wings. The wingless genes are responsible for creating patterns during embryonic development in all species, even those that normally don't have wings. This fly mutant phenotype was first identified by R. P. Sharma in New Delhi, India, in 1973. Many genetic researchers have studied wingless over the years, and found the wingless protein is a signal for other cells to do something, for example to divide a particular way.

Xenopus tropicalis: is a clawed frog native of South Africa. Scientists studying Xenopus first discovered how the basic embryonic cell cycle was regulated. Xenopus became famous due to the discovery by John Gordon and Nick Hopwood in 1958. They studied what happened when they injected the urine of possibly pregnant women under the skin of the frog: If the woman was pregnant, these frogs laid eggs; if the woman was not pregnant then the frogs did not lay eggs. These experiments led to our current day pregnancy tests. John Gordon, and also Shinya Yamanaka won the Nobel Prize in 2012, for showing that the nucleus from adult cells can be reprogrammed to become stem cells.

Yeast: or *Saccharomyces cerevisiae*, also known as baker's yeast, is a fungus used to make bread or alcoholic beverages. It is a popular model organism for studying genetics and molecular biology, particularly the cell cycle. These fungi reproduce both asexually and also by budding, which we show in this coloring book.

Zea mays: or maize is a member of the grass family that produces a cereal grain first grown by Indigenous people of Central and South America, India and China. Maize is one of the most important agricultural crops in the world. Maize was used by Barbara McClintock to study inheritance by chromosomes and meiosis, a cellular event where chromosomes exchange information. She discovered the telomere and centromere, regions on the chromosome important for preserving information each generation. She won the Nobel Prize in 1983, for her discovery of jumping genes or transposons, which she found to turn on and off the colors of the maize seeds or kernels, in the cobs of what is known as "Indian corn" or Flint corn.

References

Yourgenome.org. (2020, May 26). Retrieved June 29, 2020, from https://www.yourgenome.org/

Cool Creatures – Biomedical Beat Blog. (n.d.). Retrieved June 29, 2020, from
 https://biobeat.nigms.nih.gov/tag/cool-creatures/

Using Research Organisms to Study Health and Disease. (n.d.). Retrieved June 29, 2020, from
 https://www.nigms.nih.gov/education/fact-sheets/Pages/using-research-organisms.aspx

Facts for Kids. (n.d.). Retrieved June 29, 2020, from https://kids.kiddle.co/

The official website of the Nobel Prize. (n.d.). Retrieved June 29, 2020, from https://www.nobelprize.org/

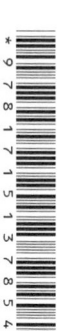

www.ingramcontent.com/pod-product-compliance
Lightning Source LLC
Chambersburg PA
CBRC100912220526
45473CB00010B/2868